Getting at the piece of the matter.

Published by The Honey Bee Company LLC
2906 Central Street • Suite 258
Evanston, IL 60201
847.864.8032

ISBN 978-1-947933-09-5

My name is

____

I'm ready to get to work!

Paige and Paxton are at Corner Pastries to order a cake for their dad's birthday.

How many cakes are there?

There are ________ cakes.

Paige and Paxton marvel at all the cool cake creations.
But that's just the beginning.

How many cakes have green frosting?

________ cakes have green frosting.

Paige and Paxton are surprised to find out that the baker is also a chemist.

Circle the biggest picture.

What's a chemist?
A chemist is a scientist.

Draw a picture of a scientist.

What is science?
Science is the way we find out how things work and why things are the way they are.

Chemistry

is the way we find out how ingredients work and why ingredients behave the way they do.

Chemists love ingredients.

Circle the one that doesn't belong.

Chemists love to mix, heat and cool all kinds of ingredients.

Color the picture.

That's why chemistry is just like baking.

Draw lines to the pictures that are the same.

Baking is combining different ingredients to get different things.

Circle the picture that's different.

It's the same with chemistry.
Everything in the world is made of the same thing: atoms.

Color the picture.

Atoms are like ingredients.

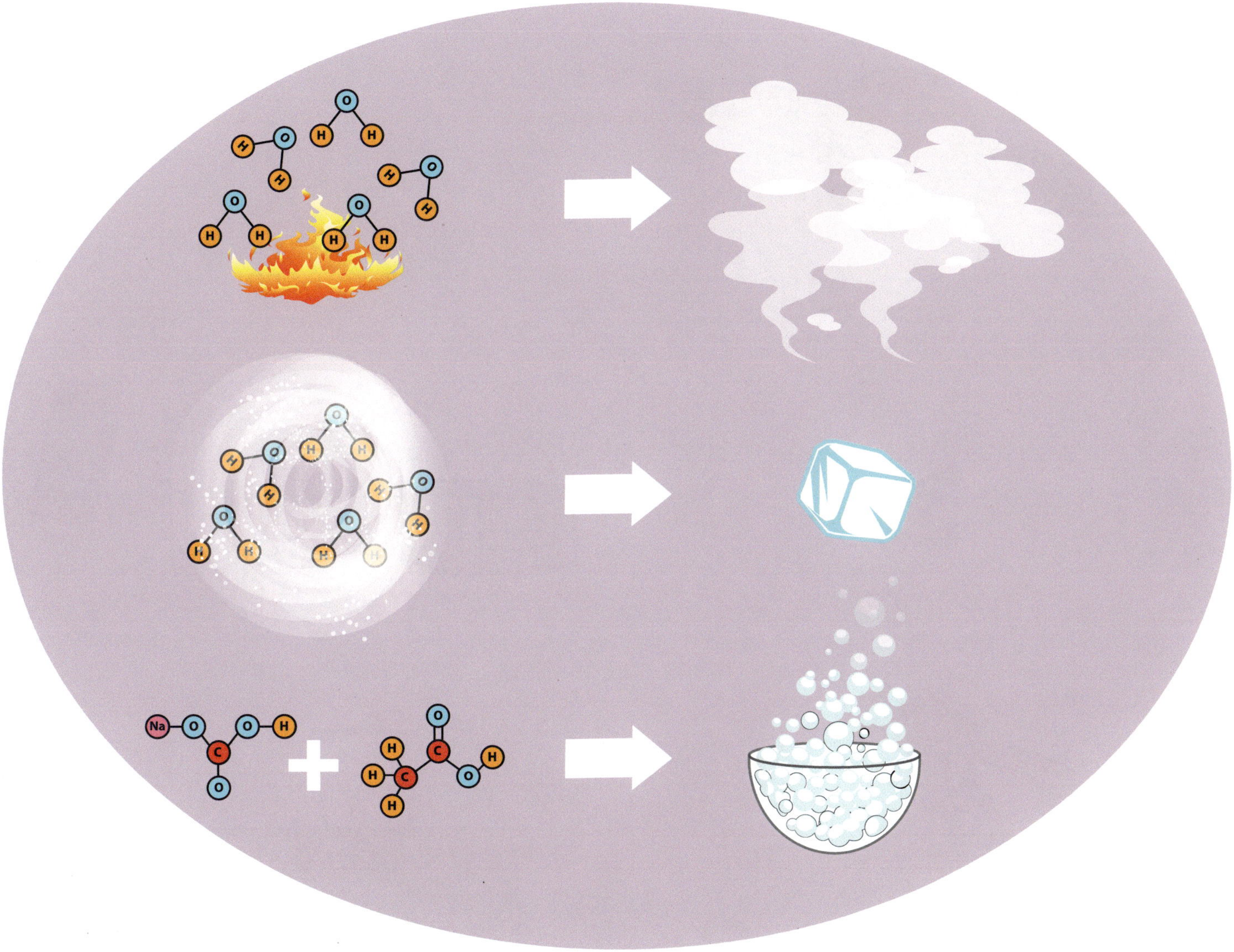

Chemistry is combining different atoms in different ways to get different things.

Atoms are tiny, tiny pieces or particles
too small to see with our own eyes.

Atoms are everywhere and in everything.

Circle all the things that are made of atoms.

How many of these things are made of atoms?

________ of these things are made of atoms.

Now draw something made of atoms.

I drew a/an ____________________________________!

Paige and Paxton are surprised to find out that there are only 118 atoms.

They're called elements,
and they're all listed on a handy chart!

Help the chemist get to her chart with all the atoms or elements.

Paige and Paxton can't believe that *everything* in the world is made from 118 different atoms.

They think it's

preposterous

What do you think preposterous means?

But that's how everything in the world is made.
Combining 118 different atoms in different ways.

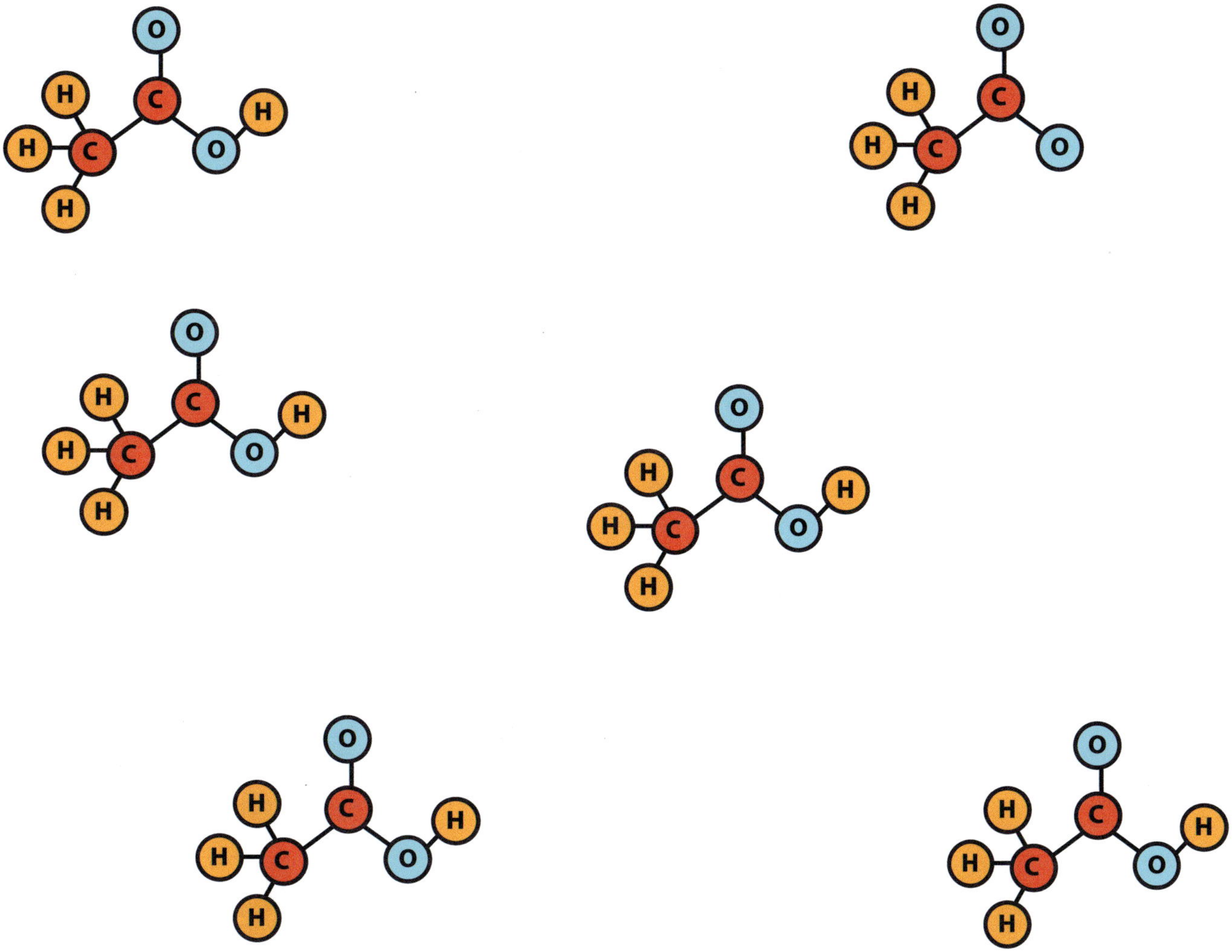

Can you find the group of atoms that's different?
Circle the group.

Paige and Paxton have learned that chemistry is the answer to many of life's most important questions.

Number the pictures from smallest to biggest.

What the world is made of…

…And the perfect cake for their dad's birthday party.

Finish

___	14	___	6	4
18	___	10	___	2

Start

Fill in the missing numbers to help
Paige and Paxton get to their dad's birthday cake.

Share something you've learned about chemistry.

Our Mission

Change the course of a nation by making STEM (Science, Technology, Engineering and Math) easy to teach and easy to learn at the very beginning of a child's educational experience

Our Vision

World Domination Through STEM

Why We Do What We Do

For Kids: 80% of adults in STEM fields had at least one parent in the field. This means a huge number of our kids are missing out on opportunities that can be life changing.

For Teachers: Classroom teachers face unprecedented challenges to meet expectations and standards. Teachers need time-saving resources that enable them to make small shifts in their classroom lessons to drive STEM interest and create a STEM mindset.

For Our Nation: We're losing the battle. Thirty years ago, 40% of the world's scientists and engineers came from the U.S. Today it's 15%. STEM fields are where the well paying jobs are—and will be—in the global economy.

Like Us On Facebook

Pin Us On Pinterest

Follow Us On LinkedIn

Follow Us On Twitter

Made in the USA
Monee, IL
05 November 2026

67380373R00019